• PRACTICAL FISHKEEPING •

# FILTRATION & WATER QUALITY

## Dr. Peter Burgess

RINGPRESS

# ABOUT THE AUTHOR

Dr. Peter Burgess BSc, MSc, MPhil, PhD is an experienced aquarium hobbyist and fish health scientist, having researched fish diseases for his MSc and PhD degrees. He is a visiting lecturer in Aquarium Sciences at the University of Plymouth, England, and edits the international science journal *Aquarium Sciences and Conservation*.

Among his other roles, Dr. Burgess is Senior Consultant to Aquarian ® fishfoods, and provides a fish health advisory service to ornamental fish traders and fish farms worldwide. He is a regular contributor to *Practical Fishkeeping* and other aquarium hobby magazines.

Commercial products shown in this book are for illustrative purposes only and are not necessarily endorsed by the author.

*Photography:* Dr. Peter Burgess (pages 5, 6, 14, 17, 25, 34, 37, 53) Aquarian ® (pages 27, 39, 48), Mary Bailey (page 119), and Keith Allison, and courtesy of Tetra UK.
*Line Drawings:* Viv Rainsbury
*Picture editor:* Claire Horton-Bussey
*Design:* Rob Benson

Published by Ringpress Books,
a division of Interpet Publishing,
Vincent Lane, Dorking, Surrey, RH4 3YX, UK
Tel: 01306 873822   Fax: 01306 876712
email: sales@interpet.co.uk

First published 2002
© 2002 Ringpress Books. All rights reserved

**ISBN 1 86054 261 1**

Printed and bound in Hong Kong through
Printworks International Ltd.

10 9 8 7 6 5 4 3 2 1

# CONTENTS

# CHAPTER 1

## UNDERSTANDING WATER

The quality of aquarium water has a profound effect on the well-being of the fish we keep. Fish that are kept under optimum water conditions will have better colour and growth, and are more likely to breed. Most importantly, they will be less likely to get ill. Indeed, the vast majority of aquarium fish illnesses are caused, either directly or indirectly, by unsuitable water conditions.

The key to successful fishkeeping therefore lies in maintaining good-quality water. The purpose of this book is to explain how this goal can be achieved.

### THE ELEMENTS

"Water conditions" is a term used to describe the various physical and chemical properties of a particular body of water. These include its temperature, oxygen content, hardness, and its acidity or alkalinity (pH). Each of these conditions is discussed in Chapter 2.

Unfortunately, it is not possible to provide a set of water conditions that will suit all fish. This is because the hundreds of fish species that are kept in aquariums originate from different regions of the world, and from different aquatic environments, ranging from rivers, streams and lakes, to ponds, muddy pools and saline lagoons. These environments vary in their water chemistry, temperature, and other factors (e.g. water movement).

# NATURAL HABITATS

*A fish generally fares best if given similar water conditions to those of its natural habitat.*

A spring-fed river in Mexico. The water is about 24°C (75°F) and slightly alkaline. Livebearers, cichlids and *Astyanax* tetras are found here.

A static water stream near the coast of Malaysia. The water here is slightly salty (brackish), and is inhabited by elongate live-bearing fish known as half-beaks.

These South American tetras and catfish prefer soft, acid water.

Each species of fish will have adapted, to a varying extent, to the particular water conditions in which it evolved in the wild. As a result, some fish do not adapt well to water conditions that are different to those in its wild habitat. Fortunately, many popular aquarium fish are more adaptable, within limits.

## WATER QUALITY

In aquarium parlance, the quality of water refers to its overall suitability for keeping fish, in particular its purity from pollutants.

There are many factors which can lead to water-pollution, resulting in a sub-standard environment for fish. The water may be polluted at source – for example, if water used to fill an aquarium has been collected from a contaminated supply.

However, the usual cause of pollution in an aquarium is the accumulation of fish wastes and other decaying matter, as a result of poor aquarium management.

Common mistakes made by many novice fishkeepers (and some not-so novices!) are:

- Overfeeding
- Overcrowding
- Stocking a new aquarium too rapidly
- Failing to undertake regular partial water changes
- Failing to provide adequate filtration.

With a little knowledge, all these mistakes can be avoided, and you can create a healthy environment where your fish will thrive.

Overcrowding and overfeeding are chief causes of poor-quality water.

# CHAPTER 2

## WATER CONDITIONS

In terms of fishkeeping, the most important physical and chemical features of water are:

• Temperature
• pH
• Hardness
• Oxygen level.

These features are examined in this chapter. The subject of chemical pollutants is covered in Chapters 4 and 5.

**BUYER BEWARE**

Before buying a fish, find out about the sort of water conditions it needs. For example, some fish require very soft, acid water, while others may need to be kept at a specific temperature. You can refer to specialist aquarium fish books for this information, or you can ask the aquarium shop staff for advice.

Armed with the relevant facts, you need to work out whether the water conditions within your aquarium fall within the fish's tolerance range. If not, don't buy it!

Keeping fish in unsuitable water could well result in the fish becoming chronically stressed, leading to illness and possibly death.

## TEMPERATURE CONTROL

Fish are 'cold-blooded' (ectothermic) creatures whose body temperature closely matches that of the surrounding water. Water temperature, therefore, has a profound influence on the lives of fish, both in the wild and in captivity.

Temperature influences the rate of various bodily functions (e.g. rate of metabolism and growth rate). For many fish, either an increase or decrease in water temperature is a trigger for spawning activity.

Most tropical fish prefer a fairly constant water temperature, usually somewhere between 21 and 26°C (70 and 79°F). In an aquarium, this is achieved with a thermostatically-controlled heater.

The water temperature affects a fish's metabolism and growth rate.
Pictured: *Herotilapia multispinosa*.

Goldfish prefer cool water conditions and do not fare well in tropical aquariums. Pictured: Common Goldfish.

Goldfish, on the other hand, need a cooler water temperature, and their optimum is around 20°C (68°F).

Although goldfish will tolerate a wide temperature range, they should not be subjected to constant tropical conditions and are therefore unsuited to the tropical aquarium.

Whether you keep tropicals or goldfish, you should install a glass aquarium thermometer so you can keep a check on the water temperature.

Never expose fish to a sudden change in water temperature, as this can shock them.

## THE pH SCALE

This is the measurement of the acidity or alkalinity of water. The pH scale runs from 0 (extremely acid) to 14 (extremely alkaline). The mid-point, pH 7, is termed neutral. Generally, pH is cited to one decimal place (e.g. pH 7.3, pH 5.8, etc.).

One-tenth of a pH unit may seem insignificant, but bear in mind that pH is a logarithmic scale. For example, pH 5.0 is ten times more acid that pH 6.0. Hence, even a change of just 0.1 pH unit has a significant effect on the acidity or alkalinity of the water.

Most freshwater fish (tropicals and goldfish) require a pH within the range 6.6 to 7.8. But there are many exceptions.

For example, the cichlid fishes of Lake Tanganyika in Africa are accustomed to highly alkaline waters (pH 8.6 to 9.2) in the wild.

Conversely, certain South American tetras and cichlids inhabit soft, acid waters and many require a low pH, around 6.0-6.5.

Most freshwater fish require a pH range of 6.6-7.8, but there are exceptions. This Ram Cichlid (*Microgeophagus ramirezi*) comes from soft, acid waters and so requires a lower pH.

## FALLING pH

The pH of aquarium water tends to fall over time, due to the accumulation of organic acids that are produced by fish and other living organisms within the aquarium.

Regular partial water changes (pages 46-48) will help to prevent these acid wastes from accumulating in the aquarium and causing a pH fall.

See also Water Hardness and pH, below.

## pH SHOCK

A sudden change in pH can be highly stressful and potentially life-threatening to fish, causing the disease condition known as 'pH shock'.

If fish do have to be subjected to a change in pH, ensure this is done gradually, no more than 0.2 pH units per 24-hour period. (See Chapter Seven, Adjusting Water Conditions.)

## WATER HARDNESS

This reflects the quantity of minerals that are dissolved in water, mostly calcium and magnesium salts. Water that contains high levels of minerals is termed 'hard', whereas water having a low mineral load is 'soft'.

There are lots of ways of measuring and describing water hardness, which can be very confusing! Most aquarium test kits measure hardness as either degrees of GH (°GH) or degrees of KH (°KH). For both scales, the higher the value the harder the water.

• **GH**: This stands for General Hardenss. It is also known as total hardness. The actual scale used to measure general hardness varies from country to country, so check which units (e.g. English, German, US) your test kit uses. English degrees of GH are used in this book. Knowing the GH value of your water is useful when selecting suitable fish. Fortunately, many popular aquarium fish tolerate water hardness within the range 4 to 20° GH, but there are numerous exceptions that fare best only in softer (less than 4° GH) or harder (above 20° GH) water.

• **KH**: This is the Carbonate Hardness. (KH derives from the German word 'Karbonatharte'). The build up of acidic wastes in the aquarium, such as those excreted by the fish, will gradually reduce the KH value of the water. Hence, a fall in KH indicates that a part-water change is overdue.

### HARDNESS AND pH

Water that is hard tends to be alkaline (above pH 7.0) whereas soft water tends to be acid (below pH 7.0).

Under soft-water conditions, the pH becomes unstable and can rapidly plummet to a dangerously low level — known by fishkeepers as 'pH crash'. Hence, if you live in a very soft-water area, it is worth monitoring the hardness (specifically the carbonate hardness) as virtual depletion of hardness will herald a pH crash.

Under such circumstances, it is wise to boost the hardness level to help stabilise the pH (see Chapter 7, Adjusting Hardness, pages 58-59).

### OXYGEN LEVELS

The oxygen that is present in water is referred to as

'dissolved oxygen'. Fish require oxygen in order to breathe, just as we do. Most fish obtain oxygen directly from the water via their feathery gills, these appendages being analogous to the lungs in other vertebrates. The fish's gills have to be very efficient because water contains only meagre amounts of oxygen, around 3 to 5 per cent of that present in air.

Some species of fish are more sensitive to low-oxygen conditions than others. Fish that inhabit fast-flowing rivers, for example, are accustomed to well-oxygenated waters.

Generally, you won't need to measure the oxygen level of your aquarium water, but you should ensure that the water is adequately aerated, to maintain the dissolved oxygen level near its maximum (termed its saturation point).

Aeration can be achieved in various ways, for example by bubbling air through the water via an aquarium air pump. Some filters are air-driven in this way. Other types of filter (notably canister filters) have an integral water pump that aerates the water through surface agitation (air is mixed with the surface water) or with the aid of spray-bar or 'venturi' attachments.

This streamlined rasbora from Borneo is accustomed to well-oxygenated, fast-flowing waters.

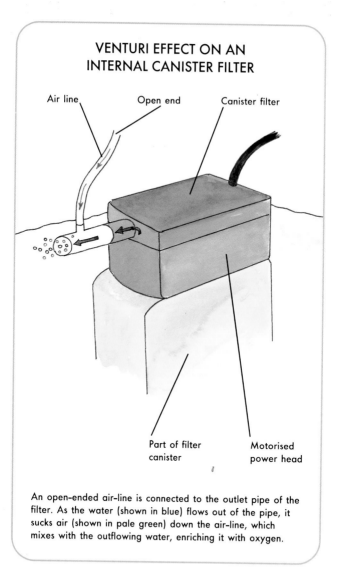

# VENTURI EFFECT ON AN INTERNAL CANISTER FILTER

Air line

Open end

Canister filter

Part of filter canister

Motorised power head

An open-ended air-line is connected to the outlet pipe of the filter. As the water (shown in blue) flows out of the pipe, it sucks air (shown in pale green) down the air-line, which mixes with the outflowing water, enriching it with oxygen.

You may see oxygenating tablets for sale. These are sometimes advocated for goldfish aquariums and bowls. In my view, these products are no substitute for proper aeration.

Aquatic plants *consume oxygen from the water at night.*

Similarly, don't rely on aquatic plants ('oxygenating weeds') to oxygenate your aquarium: they only release oxygen in the light, during photosynthesis, and actually consume oxygen from the water at night!

## OXYGEN AND WATER QUALITY

Overcrowded or very dirty aquarium conditions are likely to have low dissolved oxygen levels as a result of oxygen consumption by too many fish, or by bacteria. Goldfish bowls are particularly prone to low oxygen problems, because these receptacles are typically very small and unfiltered, and are often overcrowded.

When the dissolved oxygen in the aquarium reaches a critically low level, it is often large fish and those with a high oxygen requirement that are the first to suffer. It is not uncommon for the whole stock to be wiped out within a matter of hours as a result of plummeting oxygen levels.

Fish that are suffering from lack of oxygen may remain at the water surface and gulp or gasp. Those that have died from oxygen starvation ('hypoxia') are often found with their gills flared and mouths agape.

## TEST KITS

Test kits are available for measuring a whole range of aquarium water parameters, including pH, hardness, ammonia, nitrite and nitrate. Most give a simple colour change reaction that can be read visually.

If you don't want the expense of purchasing your own set of test kits, you may find that your aquarium store offers a water-testing service for a small fee. Give them a fresh sample of aquarium water (a cupful is plenty) in a clean glass jar or plastic container, and they will test it for you.

You can also buy electronic meters to measure certain water parameters, notably pH. Although fairly expensive, these instruments are ideal if you have many aquariums and need to do lots of water tests.

This large cichlid died from a lack of oxygen. The fish is shown belly-up — note its flared gill covers.

# CHAPTER 3

## PREPARING WATER

It is very important to select an appropriate source of water for filling the aquarium, and to ensure that the water is made safe before exposing fish to it. Generally, water can be provided from the following sources:

- Tap water
- Rain water

### TAP WATER

Most fishkeepers use water from the domestic tap to fill their aquarium. Tap water is convenient and generally of high quality, especially when supplied for drinking purposes. However, it is not 100 per cent pure (i.e. it is not pure $H_2O$) and will contain dissolved salts and other chemicals, including trace amounts of various contaminants.

'Raw' tap water can be used to clean gravel, but must be treated before being used in an aquarium.

Granite (shown above) is safe for use in both hard and soft water aquaria. Limestones (including tufa) will increase hardness and pH.

The chemical composition of tap water will vary considerably from region to region.

In some areas, tap water is very hard and alkaline because the water source has flowed over or through rocks (e.g. limestones) that leach minerals. People living in a hard-water area will notice that electric kettles fur up quickly and soap does not lather readily. In other regions, where the water has been in contact with inert rocks (e.g. granite), it will be very soft (a bar of soap goes a long way!).

The water-supply company will also chemically alter the water, by adding chlorine or chloramine in order to destroy bacteria in order to make the water safe for drinking.

These chlorine-based disinfectants, although beneficial to human health, are potentially deadly to fish and must be removed before tap water is added to a stocked aquarium (see page 20).

## PREPARING TAP WATER FOR AQUARIUM USE

The proper maintenance of an aquarium includes frequent partial water changes to dilute out accumulating wastes.

You will therefore need partly to empty and refill the aquarium with prepared tap water on a regular basis.

The step-wise procedure is as follows:

### ❶ Temperature control

Fill a clean, detergent-free plastic bucket two-thirds with tap water from the cold tap. The temperature of the water will need to be matched (to within 2°C/4°F) to that of the aquarium.

For tropical aquariums and most indoor goldfish aquariums this usually means increasing the tap water temperature by slowly mixing in some hot water from a kettle.

Aquarium heater-thermostats have control dials that can be altered to achieve the correct temperature.

Rarely, the tap water may be warmer than that in the aquarium, in which case you can float a plastic bag of ice cubes in the bucket until the required temperature is reached. Use a glass aquarium thermometer to monitor the temperature adjustment, swirling every so often to ensure a good mix.

Note: failure to match the water temperature can result in the fish suffering from cold- or heat-shock, which can stress or kill them.

## ❷ Chlorine removal

The next step is to remove chlorine-based disinfectants from the tap water. Even if you cannot smell chlorine in the tap water, it may still be present in levels that could harm your fish.

Chlorine removal, known as 'dechlorination', is simple and fast to achieve. Just add a few drops of a dechlorinator solution ('water conditioner') to the bucket of tap water, and mix well.

Dechlorinators are available under various brand names from the aquarium store. Choose a product that also eliminates the more persistent form of chlorine known as chloramine. Read the manufacturer's instructions regarding how much to add.

Check that your dechlorinator solution also removes chloramine from your water.

Siphon the water in carefully to avoid disturbing the aquarium set-up.

## ❸ Filling the aquarium

You can now add the prepared tap water to your aquarium. Pour in (or siphon in) the water slowly to minimise disturbance to the gravel, plants and fish.

Note: if you have adjusted your aquarium water conditions (e.g. raised or lowered the pH or hardness, etc.), then you must similarly pre-adjust the new tap water *before* adding it to the aquarium.

Never make these adjustments afterwards, as this can stress the fish (see Chapter 7, Altering Water Conditions).

## RAIN WATER

Fishkeepers who live in hard-water areas sometimes mix rain water with tap water to reduce the hardness (and pH). This is handy for creating soft, acid-water conditions for keeping and breeding certain types of fish (e.g. discus fish and some tetras).

## AQUARIUM BUCKETS

Plastic buckets (with a capacity of 1 to 2 gallons/4 to 9 litres) are ideal for emptying and filling aquariums, and for washing filter media, gravel, décor, and other aquarium materials, so it is worth buying at least two buckets at the very outset.

Ideally, choose buckets that have volume graduations etched on the side wall: this is handy when dispensing water conditioners and other chemical additives that need to be dosed correctly.

Never fill your designated aquarium buckets with bleaches, soaps, detergents, or other chemicals, as even traces of these cleaning agents are deadly to fish.

Tip: use an indelible marker pen to clearly label the outside of the bucket 'For aquarium use only' – this will prevent the buckets accidentally being used for domestic household cleaning chores.

However, there is a danger. Although rainwater is widely perceived as a very pure form of water, it is not always safe for fish.

Virgin rain-water is very soft and moderately acid, about pH 5.6. In contrast, rain-water that has been collected in urban areas may be very acid (as low as pH 4.0) due to contamination with industrial pollutants (the so-called 'acid rain' phenomenon).

Because of this, it is wise to seek advice from experienced fishkeepers living in your neighbourhood in order to check the suitability of locally collected rain water for aquarium use.

Never use rain water that has been collected from a lead roof, or that has been stored in a metal water container, as metal contamination may poison fish.

Even if considered safe, rain water should always be tested for pH, and adjusted where necessary, before adding to the aquarium. This avoids causing the fish 'acid shock'.

As a general rule, use rain water only where absolutely necessary.

## WATER TO AVOID

The fishkeeper would be advised to avoid using the following water sources.

## NATURAL WATER BODIES

Do not use water taken directly from lakes, rivers, ponds or other natural water bodies, as these sources may harbour fish diseases and pests (e.g. leeches). Similarly, never fill an aquarium with water from an ornamental fish pond.

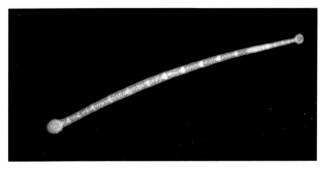

Using water from rivers, ponds, or other natural sources, can introduce parasites, such as this fish leech, to your aquarium.

Pond water may not be safe for use in aquariums.

## BOTTLED WATER

Bottled water (e.g. spring water or mineral water) is not recommended for aquarium use. Apart from being an extremely expensive way to fill an aquarium, the high mineral content of some of these products makes them unsuitable for keeping most types of fish.

# CHAPTER
# 4

## WATER FILTRATION

The fish within your aquarium are constantly producing solid and liquid wastes. Uneaten food and decaying plant leaves will also contribute to the amounts of wastes generated. In the absence of any waste-processing system, these wastes will build up in the aquarium, eventually polluting the gravel and water and poisoning the fish.

In rivers and lakes, the fish's wastes are flushed away or broken down and recycled by micro-organisms (notably bacteria) and aquatic plants. However, within the confines of an aquarium there is no constant flushing system, and the very high density of fish (compared to that in the wild) means that waste production usually far exceeds waste-processing.

The role of the aquarium filter is to boost waste-removal, making the aquarium water cleaner and safer for your fish.

### FILTER FUNCTIONS

Many modern aquarium filters perform three major functions:

- Remove solid particles, such as fish faeces, uneaten fish food, and suspended particles
- Break down ammonia fish wastes
- Help to aerate (oxygenate) the water.

Ammonia cannot be seen or smelled, but it can be deadly to fish, so tests must be performed regularly.

The filter is an essential piece of equipment for any tropical or goldfish aquarium. An aquarium filter need not be expensive, and most models are very economical to run and simple to maintain.

The most important function of a filter is to break down toxic ammonia. Fish produce two main types of waste matter: solid wastes (faeces) and liquid wastes (notably urine and ammonia). Of these, ammonia is very toxic to fish, which is why they excrete it from their bodies into the water, via their gills.

Fish that are exposed to toxic levels of ammonia in the water may suffer damage to their gills, skin and nervous system. At higher levels, ammonia is lethal. (See box-out, page 30, for recommended maximum ammonia levels.)

In addition to that excreted by the fish, ammonia is also generated from decomposing organic matter, such as uneaten fish food and rotting plant leaves.

You won't be able to see ammonia in your aquarium water, or even smell it. Hence, even crystal-clear water may contain deadly levels of ammonia.

Only by performing an ammonia test (see page 60) will you know if this dangerous poison is present.

## AMMONIA-EATING BACTERIA

In natural bodies of water, the ammonia wastes are broken down by certain types of bacteria, known as nitrifying bacteria. These bacteria gain energy from this breakdown process, and in doing so convert the ammonia into nitrite, which is far less toxic to fish.

Another group of nitrifying bacteria utilise nitrite for energy and break it down to nitrate, which is only mildly toxic to fish. In nature,

Using a test kit is the only way of knowing if ammonia is present.

the nitrate is removed from the water by aquatic plants or is further broken down by other sorts of bacteria.

These beneficial nitrifying bacteria also occur in the aquarium, but usually in insufficient numbers to cope with all the fish's ammonia wastes. This is where the filter comes to the rescue, by boosting the populations of nitrifying bacteria within the aquarium.

Most aquarium filters are designed to provide ideal conditions for the establishment of vast numbers of nitrifying bacteria within the filter medium. Such bacteria-laden filters are described as 'microbiological filters', 'biological filters', or simply 'bio-filters', and there are many types on the market.

## TYPES OF BIOLOGICAL FILTER

Walk into any aquatic store and you may be overwhelmed by the vast array of aquarium filters on sale. They vary considerably in terms of size, complexity and price. The following information will hopefully allow you to narrow down your choice.

# GENERATION OF TOXIC AMMONIA IN AN AQUARIUM AND ITS BREAKDOWN BY FILTER BACTERIA

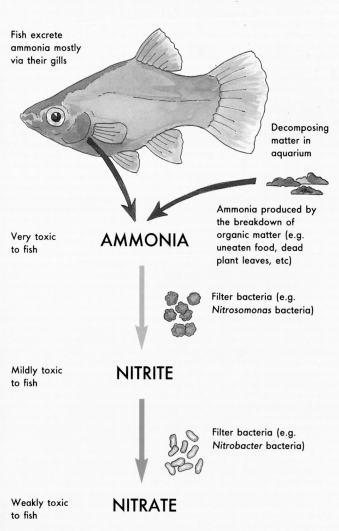

Fish excrete ammonia mostly via their gills

Decomposing matter in aquarium

Very toxic to fish

**AMMONIA**

Ammonia produced by the breakdown of organic matter (e.g. uneaten food, dead plant leaves, etc)

Filter bacteria (e.g. *Nitrosomonas* bacteria)

Mildly toxic to fish

**NITRITE**

Filter bacteria (e.g. *Nitrobacter* bacteria)

Weakly toxic to fish

**NITRATE**

The process by which bacteria breaks down ammonia to nitrate is part of a natural recycling phenomenon known as the Nitrogen Cycle.

## SAFE LEVELS OF AMMONIA, NITRITE AND NITRATE

Ideally, the levels of ammonia and nitrite should be zero (i.e. undetectable by aquarium test kits) at all times. This indicates that the biological filter is functioning properly and coping with all the fish's ammonia wastes.

### AMMONIA
Maximum recommended level: 0.2 mg/L*.
Note: the toxicity of ammonia increases with pH and with water temperature.

### NITRITE
Maximum recommended level: 0.5 mg/L*.

### NITRATE
Ideally, keep below 70 mg/L.

*If your aquarium water is dangerously high in ammonia or nitrite, then perform a largish partial water change without delay.

For example, a 50 per cent water change will halve the level of ammonia/nitrite.

Remember to dechlorinate the replacement water and adjust its temperature before refilling the aquarium (see Chapter 3, Preparing Water).

Further partial water changes may be required to maintain the ammonia/nitrite at a safe level until the underlying cause is put right.

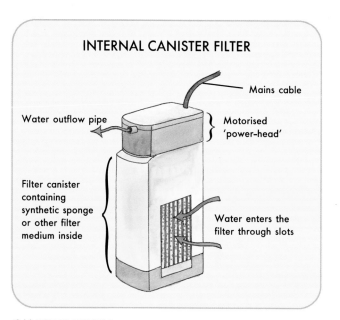

## INTERNAL CANISTER FILTER

Mains cable

Water outflow pipe

Motorised 'power-head'

Filter canister containing synthetic sponge or other filter medium inside

Water enters the filter through slots

## CANISTER FILTERS

These motor-driven filters are the most popular form of biological filter for the home aquarium. Numerous models are on the market to suit a wide range of aquarium sizes. Smaller ones tend to be of the submersible type and contain one or more synthetic sponge (foam) cartridges. (Warning: always double check that the filter is fully submersible before lowering it into water!) Most large models, on the other hand, are housed externally so they don't occupy valuable aquarium space. The only caution with outside filters is the risk of leaks with possible damage to carpets etc.

Note: the filter volume (i.e. volume occupied by the sponge or other filter medium) is just as important as its flow rate. A high-output filter that has only a tiny filter volume is unable to accommodate many filter bacteria and therefore won't be very efficient. It will also clog up quickly.

## UNDERGRAVEL (UG) FILTERS

As their name suggests, undergravel filters are positioned directly on the base of the tank, and covered with gravel. Hence they must be installed before the gravel, rocks and water are added.

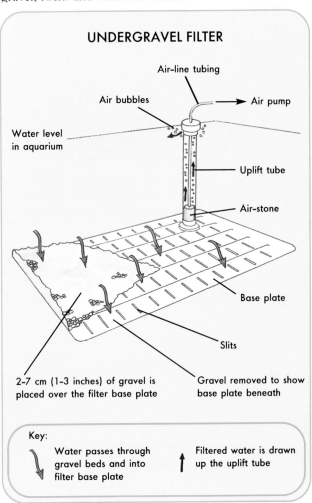

**UNDERGRAVEL FILTER**

Air-line tubing

Air bubbles

Air pump

Water level in aquarium

Uplift tube

Air-stone

Base plate

Slits

2-7 cm (1-3 inches) of gravel is placed over the filter base plate

Gravel removed to show base plate beneath

Key:

↓ Water passes through gravel beds and into filter base plate

↑ Filtered water is drawn up the uplift tube

Most models are constructed of plastic in the form of pipes or plates and are covered in numerous slits or holes that enable the passage of water.

The majority of undergravel filters are air-driven (so you will need an air pump and some airline tubing to

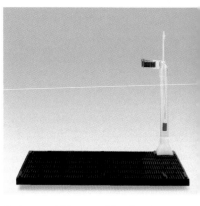

Undergravel filters work by drawing water through the gravel.

operate them), but some models can be fitted with a motorised pump ('power-head') to give higher flow rates. Undergravel filters work by drawing water through the gravel. Hence the gravel serves as the filter medium within which the nitrifying bacteria live.

You should have about 5 cm (2 inches) of gravel covering an undergravel filter. Do not use very fine gravel or sand as these may block the filter slits or holes and prevent the filter from functioning properly.

Some tropical fish, such as khuli loaches and certain catfish, have a tendency to swim down the uplift tubes of undergravel filters, but usually find their way back out!

Khuli loaches (*Pangio* species) are notorious for swimming into the tubes of undergravel filters!

## AIR-DRIVEN SPONGE FILTERS

These simple forms of bio-filter are only suitable for small aquariums (e.g. up to 25 litres/5 gallons), particularly fry-rearing systems (see Chapter 6, Aquariums for Special Needs).

A separate air pump is needed to drive them. An air uplift device draws the water through the synthetic sponge cartridge which harbours the nitrifying bacteria. The sponge is usually exposed and not enclosed within a housing. Air-driven sponge filters are submerged in the water and attached to the aquarium glass using suction caps. Most models allow easy removal of the sponge cartridge for routine cleaning without having to detach the rest of the filter.

Air-driven sponge filter. The air-line (from an air-pump) is fitted to the small, vertical tube. Aerated water passes out of the larger U-bend tube.

## BOX FILTERS

These air-driven filters were one of the earliest forms of aquarium filter but are no longer popular. They are generally suitable for small aquariums (up to about 25 litres/5 gallons). Most are made of clear plastic and are box- or cylinder-shaped. A separate air pump is needed to drive them.

An air uplift device draws water through slits in the top or sides of the filter where it then passes through one or more different filter media: typically synthetic wool and activated carbon.

A box-type filter requires a separate air pump to power it.

The box filter is positioned on the gravel, ideally at the back of the tank or behind large rocks, out of sight. They are not simple to maintain as the whole box filter unit has to be lifted out of the water and partly dismantled in order to clean the filter medium.

## MULTI-FUNCTION FILTERS

Most biological filters also work mechanically by entrapping particulate matter. Similarly, some aquarium filters, notably box and canister models, can be filled with chemical filtering agents (see box out, page 36).

Some canister filters are designed to accommodate a heater-thermostat. This is convenient and tidy, but heating problems can occur if the filter stops or becomes blocked for any reason.

## MATURING A NEW BIO-FILTER

Your new bio-filter will initially be devoid of any bacteria and so will have no biological filtering activity. It usually takes four weeks or more for the filter to become fully colonised with bacteria, at which point it is said to be 'mature'. Fortunately, you don't have to seed it with bacteria yourself for they will already be

# OTHER FORMS OF FILTRATION

## MECHANICAL FILTRATION
This is the physical removal of particulate matter such as suspended solids, plant debris, fish faeces, uneaten food and the like. Some canister filters incorporate a pre-filter (often a filter pad) which removes particles that might otherwise clog the bio-filter. Unlike a biological filter medium, the mechanical pre-filter must be washed thoroughly every week and fully replaced every so often.

## CHEMICAL FILTRATION
You can buy special filter media to remove certain chemicals from the water.

A common chemical filter medium is activated carbon which adsorbs organic substances and helps keeps the water clear. Activated carbon is usually sold as granules or specially produced pads. This medium must be replaced regularly.

Another chemical filtering medium is Zeolite that is used to remove toxic ammonia from the water (e.g. in situations where the bio-filter is not working properly). Zeolite is effective only in fresh water. It is useful for quarantine systems (see pages 53-54).

Note: activated carbon and certain other chemical filtering media should be removed before using chemical disease treatments in an aquarium. Failure to do so may result in the treatment being filtered out of the water before it has had time to work.

It takes time for filter media to mature. Pictured: the progressive dirtying of the sponge cartridge filter from new (left) to recently rinsed (middle) and very dirty (right).

present in the aquarium water and will colonise the filter by themselves.

There is no way of directly telling when your filter is mature – you cannot see the filter bacteria! But you can get a good clue by regularly monitoring (every few days) the ammonia and nitrite levels of the aquarium water: their levels should rise and then fall as the filter matures. When both have fallen to zero, this suggests filter maturation – i.e. the filter has sufficient bacteria to cope with all the fish's ammonia wastes.

## NEW TANK SYNDROME

It is extremely important not to add too many fish to a new aquarium. Just add a couple of fish initially, and then a few more two to three weeks later, gradually building up the numbers over time. If you put in too many fish at once, then the maturing filter may be unable to cope with their wastes, resulting in dangerously high levels of ammonia and/or nitrite in the water. This problem is widely known as 'New Tank Syndrome'. Perhaps a more appropriate name would be 'New Filter Syndrome'.

Novice fishkeepers who are too eager and impatient to increase their fish numbers gradually risk losing their fish to ammonia or nitrite poisoning. In despair, many give up the hobby there and then.

## SPEEDING UP FILTER MATURATION

For most situations you can let the filter mature in its own time. But sometimes it may help to speed up this maturation process, for example if the new filter has to cope with a lot of fish.

Methods for accelerating filter maturation time are:

1. Use a bacteria starter culture. You can purchase liquid or dry (dormant) cultures of bacteria from the aquarium store to help speed up filter maturation. Get advice before choosing a starter culture product: some are of dubious efficiency.

2. Use some medium from a mature filter to seed the new one. The donor filter medium must not be taken from an aquarium with a recent history of disease, otherwise it could harbour fish pathogens!

3. Temporarily (i.e two weeks) 'park' the new filter in an established aquarium (disease-free!) so it quickly acquires nitrifying bacteria.

Filter bacteria require a constant flow of water to keep them — and, in turn, the fish — healthy. Pictured: Angelfish, *Pterophyllum scalare*.

## CHANGING A FILTER

If you plan to change your filter, perhaps because you wish to buy a new model or a different type, then it is important to run both the old and new filter simultaneously in the aquarium for a couple of months. This allows time for the new filter to mature, thus avoiding 'New Filter Syndrome'.

## LOOKING AFTER YOUR BIOLOGICAL FILTER

Consider your biological filter as a waste-processing factory where the workers are microscopic bacteria. As with any factory workforce, the filter bacteria require certain working conditions in order to do their job.

Basically, they need a surface on which to attach and a continuous flow of water passing over them.

The greater the bacteria workforce the factory can accommodate, the more efficient the factory will be. Hence, most biological filters are filled with a porous medium (such as synthetic sponge) that provide a huge surface area for the bacteria to colonise.

In the case of undergravel filters, it is the overlying particles of aquarium gravel that provide such a surface for bacterial colonisation.

The filter bacteria that break down toxic ammonia wastes are 'aerobic', meaning they must have oxygenated water passing over them at all times in order to survive and function. Hence, a biological filter has to be kept running continuously.

If the filter fails or is switched off for more than a couple of hours then the bacteria will begin to die of oxygen starvation.

The filter bacteria's other requirement is food. The toxic ammonia wastes produced by fish are 'eaten' by the bacteria and converted via a two-stage process into relatively harmless nitrate (see the diagram on page 29). So the constant flow of water across the filter medium supplies the resident bacteria work-force with both their oxygen and food needs.

Keeping the filter factory bacteria alive is the key to proper filter management, and in turn is fundamental for achieving optimal water conditions for the fish's health.

## GOLDEN RULES

There are a number of all-important rules to observe in order to keep your filter functioning correctly.

### Never switch off the filter for any length of time

If the filter is off for more than a couple of hours, the filter bacteria will begin to die through lack of oxygen. (This is something to consider before siting an aquarium in a bedroom: if you are a light sleeper then the noise of some filters, especially air-driven models, may keep you awake at night!)

Check that the filter re-starts after being switched off (or after a power cut). If there is gunge in the filter impellor, it may fail to re-start.

 ### Ensure the flow of water (or air) through the filter is not impeded

If the filter is overdue for a clean, it can be blocked with gunge. Canister filters may stop working if the external moving parts (e.g. the impeller) are gunged up, so will need occasional cleaning.

Warning: never attempt to open the sealed motor housing. See manufacturer's instructions for routine servicing.

 ### Take special care when washing the filter medium

Do not wash the filter medium under a tap. Chlorine in tap water will kill the filter bacteria.

Also, don't wash the filter medium with soap or detergents, as these cleaning agents are highly toxic to fish, causing potentially fatal gill and skin damage.

Overwashing will remove too many bacteria, resulting in the filter factory being less efficient due to a reduced work-force. (It takes time for the filter bacteria population to multiply and recover.)

If too few bacteria remain after washing, they may be unable to cope with the fish's ammonia wastes, resulting in poisonous ammonia accumulating in the aquarium water.

The safest way to wash the filter medium is to *gently* hand rinse it in a bucket of freshly-drawn aquarium water (the dirty water is then discarded – a good garden fertiliser!).

If the filter medium is in sections (e.g. two or more sponge cartridges, or a 'sandwich' of several filter mats), then wash only half at a time. Wash the other

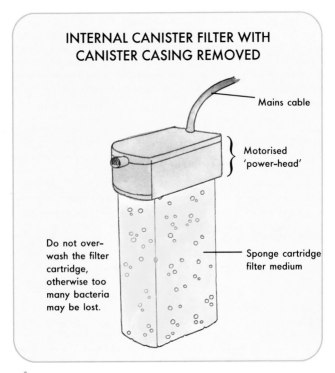

**INTERNAL CANISTER FILTER WITH CANISTER CASING REMOVED**

Mains cable

Motorised 'power-head'

Do not over-wash the filter cartridge, otherwise too many bacteria may be lost.

Sponge cartridge filter medium

half next time, and alternate. This ensures that the filter is left with at least 50 per cent of its bacteria loading.

---

### TWIN FILTERS

**If you opt for canister filtration, consider buying two small filters instead of a single large one. This is a safeguard against one unit failing or blocking, and leaving the aquarium without filtration. Having two filters means you can wash them alternately, ensuring that one filter will always have a full complement of bacteria.**

---

 **Certain disease medications may destroy filter bacteria**

Filter-harming medications include:
- Formalin
- Various dyes (e.g. gentian violet, malachite green, methylene blue, *para*-rosaniline and amino-acridine compounds)
- Some antibiotics (e.g. erythromycin).

Before using a chemical disease remedy, always check the manufacturer's instructions regarding any possible harm to the filter bacteria. Some fishkeepers remove the bio-filter during treatment, but there is a good argument for leaving it intact, for two reasons:

**1** Some of the filter bacteria may survive the medication and hence provide some filtering function until their numbers recover.

**2** Any pathogens or parasites 'hiding' in the filter will be exposed to the medication and be destroyed.

Before dosing the aquarium with filter-damaging

**Filtration And Water Quality**

disease remedies, some fishkeepers temporarily park the bio-filter in another aquarium, to avoid the filter bacteria from being destroyed. This practice is very unwise, as the filter may transmit the disease to the host aquarium!

## WHAT IF MY BIO-FILTER STOPS?
A prolonged power cut, or accidentally switching off the filter (and forgetting to reconnect it!) may cause the filter bacteria to die.

If the flow of water to the filter has ceased for just a few hours, then some of the bacteria may still be alive and hence will provide some (albeit reduced) bio-filtering function until such time as their numbers return to full-strength.

It is always a good idea to do a reasonably large (e.g. 30-50 per cent) water change after a prolonged filter stoppage (of six hours or more), and to monitor the ammonia and nitrite levels for a week or two afterwards, just in case the filter bacteria population is reduced and cannot cope with the fish's wastes.

If the filter has stopped for several hours or more then the conditions within the filter chamber may have become anaerobic (i.e. devoid of oxygen), and all or most of the nitrifying bacteria will have died.
In such circumstances:

- Don't restart the filter or it may pump out poisons (such as hydrogen sulphide) into the aquarium that could kill the fish (if the filter medium smells of rotten eggs this indicates the presence of poisonous hydrogen sulphide).
- Dismantle the filter, remove and discard (or

Switching on a filter that has been off for several hours can be lethal to fish – unless the filter medium is cleaned.

thoroughly wash) the filter medium, and then reassemble.

• The reassembled filter will have to go through a maturation process, so you must monitor for ammonia and nitrite as you would for a new filter (see pages 35-37).

# CHAPTER
# 5

## MAINTAINING THE ENVIRONMENT

Even though you have a filter installed, you still need to undertake routine tasks in order to maintain good water conditions in your aquarium.

> **Biological filtration + regular partial water changes = good aquarium management!**

## THE SOLUTION TO POLLUTION IS DILUTION!

As stated previously, the filter bacteria convert ammonia into nitrite and then nitrate. Under aquarium conditions, the nitrate (and certain other organic waste substances) are not further broken down to any extent and so will accumulate in the water. High levels of nitrate (say, above 70 mg/L) can stress fish and reduce their growth and performance.

The best and easiest way of controlling the levels of nitrate and other accumulating wastes is through partial water changes. Live aquatic plants will help by consuming some nitrate, but the level of planting in most aquariums is not sufficient to cope with all the nitrate produced.

Beware of products on the market that claim to dispense with the need for regular water changes. In my experience, there is no substitute for a partial water change every so often.

Performing regular, partial water changes is the most effective way of controlling waste and pollution.

## PARTIAL WATER CHANGES

The question of how much water to change, and how often, will depend partly on the size of aquarium, the density of the fish, and the efficiency of the filtration. The smaller the aquarium water volume or the greater the stocking density, the more frequent the partial water changes required.

Assuming you have a standard-sized aquarium (around 45-90 litres/10-20 gallons) that is not overcrowded with fish, then aim to change about one-fifth to one-quarter of the aquarium water every two weeks.

Avoid undertaking very large water changes (above 50 per cent) unless absolutely necessary: these can stress the fish.

The easiest and best way of removing the water is by siphoning it into a clean plastic bucket. Take care not to suck up any small fish! You can combine the partial water change with siphoning over the gravel, to remove any excess trapped dirt. Special gravel cleaners are available for attaching to the siphon tube to enable this.

Note: if you have installed an undergravel filter, then bear in mind that the gravel serves as the filter medium – don't clean it too thoroughly or you will remove too many nitrifying bacteria.

Use a clean plastic bucket to refill the aquarium with water. You can either gently pour in the new water or use a siphon tube.

Remember to dechlorinate and adjust the temperature of the tap-water (see Chapter 3) *before* refilling the aquarium.

## OVERCROWDING

An overstocked aquarium is a recipe for disaster. Excessive numbers of fish will place a high demand on the limited amounts of oxygen present in the water. Also, the large quantities of solid and liquid wastes

Overcrowding is a major contributing factor to poor-quality water, so the aquarium should not be overstocked.

produced by so many fish could overwhelm the bio-filter, causing toxic ammonia or nitrite to accumulate in the water.

Sooner or later, the overcrowded aquarium will reach a crisis point, and the fish may die of oxygen starvation and/or water pollution.

## OVERFEEDING

Overfeeding is a common mistake made by novice fishkeepers. As a general rule, feed little and often, say two to three times each day, giving only as much food as the fish will consume within about five to ten minutes. Any uneaten food will decompose and pollute the water.

If you have accidentally overfed, then quickly net or siphon out the excess.

Overfeeding is another cause of tank pollution.
Pictured: Clown Barbs (*Puntius everetti*) feeding on tablet food.

Fish should be fed little and often, and uneaten food should be removed immediately.

## TOP TIPS

1. Install a biological filter and maintain it properly.

2. Do not overcrowd the aquarium with too many fish.

3. Do not overfeed.

4. Perform regular partial water changes.

5. Keep the gravel clean of excess trapped dirt.

6. Monitor water parameters as necessary.

# CHAPTER 6

## AQUARIUMS FOR SPECIAL NEEDS

### THE GOLDFISH AQUARIUM

It is a cruel myth to suggest that goldfish can be kept successfully in tiny bowls and without filtration. Note the word 'successfully'. Many do survive, but their quality of life is likely to be poor, and premature death from water pollution is common. This partly explains why few aquarium goldfish reach their expected life span of 10-15 or more years.

Goldfish should never be kept in bowls — they require large aquariums that are adequately filtered.

All goldfish should be housed in large aquariums that are bio-filtered. The best type of filter for a goldfish aquarium is the internal canister type (see page 31).

Avoid purchasing small 'complete' goldfish aquariums that come with tiny, air-operated undergravel filters – they may have inadequate filtering capacity.

## THE FRY AQUARIUM

Many fishkeepers discover the satisfaction of breeding and rearing their fish. One major challenge to raising fry (in addition to providing suitable fry foods) is the provision of good water quality.

Fry are far more susceptible to poor water conditions than are adult fish and it is very easy to pollute the fry tank because of the need to give frequent feeds.

The problem with conventional canister filters is the typical strong suction of the inflow vents, which can draw in very small fry and squash larger ones between the slits.

Undergravel filters are a viable alternative. However, many expert fish breeders prefer to have bare bottom rearing aquariums for reasons of hygiene, which obviates the use of this type of filter.

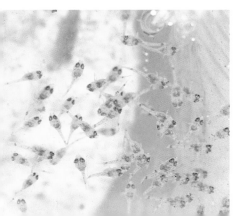

Fry are especially susceptible to poor water conditions. Pictured: cichlid fry.

Covering the filter inlet slots with plastic mesh will prevent fry from being sucked into the filter chamber.

One popular filter for rearing fry is the air-driven sponge filter (see page 34) which is ideal for a small rearing aquarium (e.g. 10 to 25 litres/2 to 5 gallons).

Another alternative is to modify a canister filter by placing some mesh across the inlet vent to reduce the risk of fry being sucked in to the filter. A couple of lengths of nylon string or rubber bands (rubber will perish over time) are used to secure the mesh in place.

Do ensure that the mesh doesn't become blocked – it will need frequent cleaning (e.g. under a running tap) otherwise the filter will be ineffective.

## THE QUARANTINE AQUARIUM

Ideally, all newly-purchased fish should be kept in isolation for 10 days minimum and up to four weeks, before placing them in the main aquarium. This will greatly reduce the risk of introducing diseases that could wipe out your existing fish stock.

The problem with using bio-filtration is that the quarantine aquarium is likely to be out of use for long periods of time (unless you are constantly buying new fish), and its bacterial load may decline in the absence of any fish to provide ammonia 'food'.

Also, certain chemicals commonly used to medicate fish can kill the filter bacteria, rendering the bio-filter ineffective and risking a high ammonia or high nitrite crisis.

Canister or box filters are good choices for use in quarantine aquariums, but neither should be relied upon to function biologically, for the reasons stated above.

To compensate for a lack of biological function, the filter should be filled with a small quantity of zeolite granules (available from the aquarium store). Zeolite is a natural compound that removes toxic ammonia produced by the fish.

Remember to replenish the zeolite periodically: it can be recharged – see manufacturer's instructions.

## LOW WATER TURBULENCE AQUARIUM

The high water output of some canister filters (especially large models) may be too powerful for certain types of fish, such as the graceful gouramis that are generally weak swimmers adapted to slow-moving or static waters.

Certain strains of fancy goldfish that have long fins and unnaturally-rounded bodies (e.g. Fantails, Moors) are also comparatively weak swimmers.

Fancy goldfish, such as the Black Moor, are weak swimmers and so need low water turbulence.

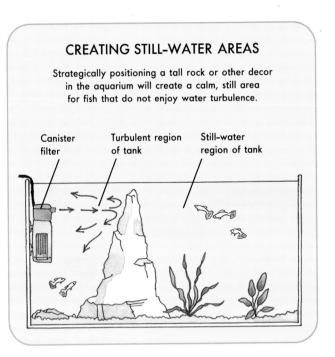

## CREATING STILL-WATER AREAS

Strategically positioning a tall rock or other decor
in the aquarium will create a calm, still area
for fish that do not enjoy water turbulence.

Canister
filter

Turbulent region
of tank

Still-water
region of tank

Such fish are at risk of being constantly buffeted and stressed if subjected to water turbulence from a high output filter. In these situations, the filter outflow can be localised by placing an upright partition close to the filter outlet. A tall rock or piece of slate works well. This gives slow-moving fish the opportunity to retreat to the quieter region of the tank, if they so desire.

Alternatively, fitting a spray-bar attachment to the filter outflow pipe will not only reduce the current, but also improve aeration.

# CHAPTER
# 7

## ADJUSTING WATER CONDITIONS

Wherever possible, avoid adjusting your water conditions. It is far simpler to select fish that are suited to your local water conditions (notably pH and hardness) than to attempt to manipulate your aquarium water chemistry to suit the fish.

If there is an aquarium shop in your local area, the chances are that it will receive the same water supply as yourself – ask the staff what types of fish fare well in the local tap water.

This Chequer Barb (*Puntius oligolepis*) will adapt to a range of pH and hardness values, within limits.

It is useful to know the pH and hardness of your local water so you can choose fish that are best suited to these conditions. In general, soft waters are acid, and hard waters alkaline.

As mentioned earlier, many popular aquarium fish will adapt to a fairly wide pH and hardness range, hence adjustments are usually unnecessary. However, adjustments may be necessary in certain situations:

 **If your water supply is very soft or very hard, or very acid or very alkaline**

Defining these limits is not straightforward. As a rough guide, it may be necessary to adjust pH if your aquarium water is above 7.8 or below 6.5.

Similarly, an adjustment may be required if the water is very hard (e.g. above 20 dH) or very soft (e.g. below 5 dH). Of course, certain types of fish will actually prefer these extremes!

See pages 12-13 for scales of hardness.

 **The fish you intend to keep have strict requirements for pH and/or hardness**

As mentioned before, some fish need hard, alkaline waters, others require soft acid conditions.

Before you actually buy any fish, ask the supplier if the particular species require special water conditions, and make sure you can meet these requirements before purchasing.

 **You need to stabilise the pH of very soft water**

This may be necessary in order to avoid a pH crash (see page 13).

## ADJUSTING pH

This can be achieved by adding a commercial pH 'buffer' that is sold in liquid or powder form specifically for aquarium use. Various buffer products are available for increasing, decreasing, or stabilising the pH.

Follow the manufacturer's instructions carefully regarding dosage and remember to undertake pH changes slowly (not more than 0.2 pH units per 24 hour period). Ensure that the pH buffer is suitable for freshwater aquariums as some are intended for marine systems only.

Note: reducing the pH of hard water is difficult as its high mineral content will 'bounce' the pH back towards alkaline. It is therefore important to soften the water (see section below) before attempting to lower its pH.

## ADJUSTING HARDNESS

Commercial salt mixes are available for increasing the water hardness of freshwater aquariums (don't confuse with aquarium sea salts sold for marine systems!).

These preparations generally contain calcium-based salt mixes. The addition of calcium-rich rocks, such as limestone, to the aquarium will facilitate a more gradual increase in water hardness.

Commercial products are available for increasing the hardness of aquarium water.

Reducing hardness is more problematic. Mixing rainwater with freshwater to give the required hardness level is one option, but rainwater has certain disadvantages (see pages 22-24).

Safer alternatives to rainwater are available, but these can be expensive:

- Distilled water: this can be purchased from pharmacists.
- Reverse Osmosis (RO) water. Some aquatic shops sell RO water, or you can buy a RO unit (which is fairly expensive) and make your own RO water at home.
- Ion-exchange (IEX) water. Special IEX resins are available for aquarium use.
  *Warning*: never use water that has been processed by a domestic water softener – it could be very harmful to the fish.

## ADJUSTING WATER TEMPERATURE

Thermostatically-controlled heaters ('heater-stats') are available for tropical aquariums. Many models are factory set at about 24°C (75°F), which is suitable for most tropical fish, so you are unlikely to need to alter them (all are adjustable, however).

For certain situations it may be desirable to adjust the aquarium water temperature. For example, certain types of tropical fish spawn in response to an increase or decrease in water temperature, hence the heater-stat in the breeding aquarium may need adjusting.

Carry out only a small adjustment at a time (especially if fish are present!) and wait a few hours for the new temperature to stabilise before making any further changes.

# APPENDIX
# 1

## TESTING WATER PARAMETERS

It can be confusing to keep a track of what needs checking when – here's a summary of the tests you should perform on your aquarium water.

### TEMPERATURE
Check the aquarium thermometer every day.

### AMMONIA AND NITRITE
Test every few days during the first four weeks or so after a new aquarium and filter has been set up.

If the results show a problem, then test every day until the conditions improve. Ideally, you should continue monitoring ammonia and nitrite levels until both have stabilised around zero.

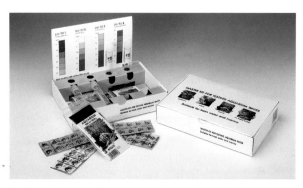

Various simple test kits are available for checking your aquarium water parameters.

Resume monitoring if the filter medium has been totally replaced or inactivated for any reason.

## NITRATE

Check the level every 4-6 weeks.

## pH

Test every month or so – more often if the aquarium water is very soft and hence prone to pH falls.

## HARDNESS

It is worth testing the hardness of your local tap water so you can select fish that suit it.

Routine testing is generally unnecessary, except if you live in a very soft water area (see Chapter 2, Hardness and pH, pages 10-13).

It will be necessary to monitor any adjustments made to your aquarium water parameters, such as when altering the pH or hardness (see Chapter 7).

# APPENDIX

## 2

## SCIENTIFIC DATA

Conversions given to two decimal places.

### LENGTH

1 metre = 100 centimetres (cm)
1 centimetre = 10 millimetres (mm)
1 mm = 1000 microns (μm)
1 centimetre = 0.39 inches (in)
1 inch = 2.54 cm

### WEIGHT

1 kilogram (kg) = 1000 grams (g)
1 gram = 1000 milligrams (mg)
1 milligram = 1000 micrograms (μg)
1 kg = 2.20 pounds (lbs)

## VOLUME

1 litre = 1000 millilitres (ml or mL) = 1000 cubic centimetres ($cm^3$)

1 litre (l or L) = 0.22 imperial gallons*

1 imperial gallon = 4.55 litres
1 US gallon = 3.79 litres

1 cubic foot of water = 6.23 imperial gallons

* In the UK the imperial system is used for gallons.

## AQUARIUM CAPACITY

To calculate aquarium volume: multiply length x height x width (in cm) to give volume in cubic centimetres ($cm^3$). Divide by 1000 to give volume in Litres.

For example, an aquarium of dimensions 90 cm length, 30 cm depth, 30 cm width has a volume of 81,000 $cm^3$. Divide by 1000 = 81 litres (= 17.8 imperial gallons).

These calculations give the maximum volume. In practice, the actual volume of water present in an aquarium set-up is typically 15-20 per cent less. This is due to the displacement of water by the rocks and gravel, plus the fact that the aquarium is generally not filled to the brim.

## TEMPERATURE

°C = degrees Centigrade (also known as Celsius)
°F = degrees Fahrenheit

Centigrade to Fahrenheit conversion
(Fahrenheit values given to
nearest degree):

| °C | °F |
|----|----|
| 0  | 32 |
| 5  | 41 |
| 10 | 50 |
| 20 | 68 |
| 21 | 70 |
| 22 | 72 |
| 23 | 73 |
| 24 | 75 |
| 25 | 77 |
| 26 | 79 |
| 27 | 81 |
| 28 | 82 |
| 29 | 84 |
| 30 | 86 |